Special Publications 74

NAVIGATING AN ACADEMIC CAREER

A Brief Guide for PhD Students, Postdocs, and New Faculty

Jeffrey J. McDonnell

This Work is a co-publication of the American Geophysical Union and John Wiley and Sons, Inc.

AGU 100
ADVANCING EARTH AND SPACE SCIENCE

WILEY

This Work is a co-publication between the American Geophysical Union and John Wiley & Sons, Inc.

This edition first published 2020 by John Wiley & Sons, Inc., 111 River Street, Hoboken, NJ 07030, USA and the American Geophysical Union, 2000 Florida Avenue, N.W., Washington, D.C. 20009

Published under the aegis of the AGU Publications Committee

Brooks Hanson, Executive Vice President, Science

Lisa Tauxe, Chair, Publications Committee

For details about the American Geophysical Union visit us at www.agu.org.

Wiley Global Headquarters

111 River Street, Hoboken, NJ 07030, USA

For details of our global editorial offices, customer services, and more information about Wiley products visit us at www.wiley.com.

Library of Congress Cataloging-in-Publication Data is available.

ISBN: 9781119642107

Cover Design: Wiley
Cover Images: © Valerio Cerantola

Set in 10/12pt Times New Roman by SPi Global, Pondicherry, India

V10015897_120219

For my students' students

Academic happiness = reality – expectations.

Anonymous

CONTENTS

PREFACE

This little book is based on a series of "Working Life" articles I wrote for *Science* magazine. Most of these came about as I reflected on my early career and the mistakes I made along the way, and on the advice I would give to my younger self if I somehow had that magical power. Many ideas too came from mentoring my own 75 or so graduate students and postdocs over 30 years and helping them launch their careers, something that was deeply rewarding for me.

My story is perhaps similar to many: I was the first in my family to go to university. But going to grad school and then navigating the transition to a faculty position left me without a frame of reference for the journey, not even a white-collar connection to link my experience with that of my family. In my early faculty years, I felt like an explorer in a foreign land, my degrees were from Canada and New Zealand, making the start of a U.S. based academic career a bit mystifying. I had no mentors. While everything turned out well in the end (I think!), I followed a path that was perhaps bumpier than most. I like to think that this gives me a useful frame of reference for telling these stories; figuring it out by myself, for better or worse, has given me perspective and empathy for the plights of others. There is no doubt that I have been exceptionally lucky and the beneficiary of the extreme kindness of senior colleagues along the way. My hope is to pay this forward here in a small way by sharing my journey in these bite-size reflections.

My goal with this small book is for it to be read in a couple of hours. While my examples come mostly from North America, I hope that many of those examples and much of the discussion transcend geography. I should state here that this book is not comprehensive. It does not get into issues of teaching, committee work, nonacademic career tracks or the killing fields of university administration.[1] Many are more qualified than I to discuss these topics, and there is a long list of recommended reading at the end of this book for those interested in these topics. This book is written from my perspective, which is limited to the many unearned advantages of my gender and nationality.[2]

This book aims to help its readers to navigate life in academia, and, like any good sailor, the student, postdoc, and new faculty member should know how to plot the course to steer. I have no special qualifications to write such a book,[3] only the experience of a sailor who has gone adrift once or twice and has been blown off course many times.

I hope my advice will help others as they launch and then navigate their own academic careers.

Jeffrey J. McDonnell
Saskatoon SK, Canada, and Corvallis OR, U.S.

ABOUT THE AUTHOR

Jeffrey J. McDonnell was born in Toronto, Canada, and has a BSc (Hon) from the University of Toronto and an MSc from Trent University in Canada. He earned his PhD in 1989 from the University of Canterbury, New Zealand where he was a Commonwealth Scholar. He then moved to the United States and was Assistant Professor at Utah State University (1989–1993), Associate and Full Professor at SUNY College of Environmental Science and Forestry (1993–1999), and Richardson Chair in Watershed Science and University Distinguished Professor at Oregon State University (1999–1912).

Since 2012, he has been Professor of Hydrology and Associate Director of the Global Institute for Water Security at the University of Saskatchewan, Canada. Jeff is an elected Fellow of the Royal Society of Canada (Canada's National Academy of Science) and Fellow of the American Geophysical Union (AGU) and the Geological Society of America.

He was the 2016 winner of the International Hydrology Prize from the International Association of Hydrological Sciences, UNESCO, and the World Meteorological Organization. He has received the Dalton Medal from the European Geophysical Union and the Birdsall-Dreiss Distinguished Lecturer

Award from the Geological Society of America. Jeff is past President of the AGU Hydrology Section, a Visiting Distinguished Professor at Tsinghua University, Beijing Forestry University, and Ludong University in China, and a Visiting Chair of Water Sciences at the University of Birmingham, UK.

Part I: The PhD Student and Postdoc Years

Only do what only you can do.

Anonymous

1

AN ACADEMIC CAREER

As an undergraduate, I worked in a professor's lab, and I saw him as a workaholic without a life. I also viewed the academic life as something I would never aspire to, given that so many of my other professors could be similarly described. Now, as a professor myself looking at 60, I wonder what I would say to my former self, how I would explain why I did it (academia, that is), and why I continue to love it?

All this self-reflection was prompted by discussions in a monthly postdoc mentoring lunch that I lead at my institute. Some postdocs openly ask, "Is it worth it," just like I did back then. They wonder why someone would subject themselves to a life so driven.

Navigating an Academic Career: A Brief Guide for PhD Students, Postdocs, and New Faculty, Special Publications 74, First Edition. Jeffrey J. McDonnell.

Admittedly, I do have a spasm of guilt when I try to "sell" the idea of university faculty track. It is certainly something to go into with eyes wide open and with as much information as one can gather.

What drives the decision to pursue an academic career? For most of us, it is great curiosity and passion for our science. After my masters program, I spent a year working in the consulting industry. I hated the structure of organizational rules and I found the repetition and narrowness of the scientific tasks boring. I spent a year after my PhD working at NASA. It was thrilling on many levels, but again, I chafed against the bureaucracy and my limited scope to study what I wanted. I simply hate rules. I love freedom.

Beyond professors' love of their science, the rejection of the conventional workplace structure is perhaps the most common reason that academics love academia, despite its adverse effect on work–life balance. My life could be described as "always working; never working," but I like to frame it and think of myself as an "*engaged workaholic*."[4]

This was different when I had young children at home. Then, it was an either-or situation, work at work or be home at home. Now, as my children reach adulthood, my work life blends into my personal life, but my wife and I try to combine my research and conference travel with add-on trips of our own. I was rarely able to do this years ago when a speedy return to parenting obligations drove my travel decisions.

Freedom, of research topics, travel destinations, is a drug. The deadline-driven life that I see my friends in the consulting industry still living (even in their later career) is mostly a self-imposed one in academia. It is that freedom to set your own schedule that is also so appealing. At the same time, however, academic work–life balance becomes most clear over the course of a career, rather than during one's first few years. There is almost no work–life balance in the pre-tenure years. Saying otherwise is simply misleading and untrue. For my first five years as an academic, I worked most nights and weekends. But, overall, and during a 30-year career, I can attest that one can indeed achieve a work–life balance to rival that of many other PhD-level occupations.

Apart from the often-discussed hubbub of faculty and committee meetings, an academic working life is the ultimate selfish pursuit: It's a curiosity-driven life, which is also addictive. Then there is the fact that even as you age, your incoming PhD students and postdocs will always be young with fresh minds. As they run with your ideas, this added inspiration is a significant bonus of the job. Plus, basking in your former students' own academic-career success is perhaps the ultimate reward. I relish attending conferences and workshops where I can watch my former students and postdocs lead the field.

But if work–life balance in academia deeply worries you, ask yourself if you would truly work 9 a.m. to 5 p.m. outside academia. As a PhD student, you are probably already wired to overachieve. If you were to work in a 9–5 environment, you might be the one to lead the union, volunteer to organize workplace events, or work nights to move up the corporate ladder. And at the end of an academic career, retirement can be like a permanent sabbatical (my plan), where unlike a PhD research career in industry, academics can retain Emeritus status with their university and play at research until the bitter end.

Ultimately, an academic job is not 9–5. But it is never, ever, boring. And, as with other professions such as emergency-department doctor, travelling salesperson, or police officer, the hours and the nature of the work are not for everyone. And that's okay because there are a multitude of things one can do with a PhD outside academia.

But this book is not about that. It's about demystifying the academic career track. Taking that track is a willing assumption of risk.

But for those who take that risk, it is often well worth it. So, in the words of David Bowie, let's now "turn and face the strange!"

...A spirit of indomitable perseverance has characterized nearly all successful scientists.

William Ian Beardmore Beveridge, 1908–2006, Professor, Cambridge University

2

TRAITS OF SUCCESSFUL SCIENTISTS

As I entered my first faculty position, I thought that the only real trait needed to be a successful scientist was academic brilliance. I believed that science was an IQ meritocracy. Wrong! I've since realized that soft skills are crucial to navigating and enjoying an academic career. I've seen tenure denials handed down in the same week that the candidate had a cover article in *Science*. I've seen the most brilliant minds struggle with the most mundane aspects of departmental life.

Now, with some gray hair, having chaired many awards committees, and knowing many members of the U.S. National Academy of Sciences and the U.K. and Canadian Royal Societies, I can identify common traits and behaviors among successful scientists. These are neither mysterious nor unfamiliar: many of them have been discussed over decades and centuries by more astute observers than I.

Navigating an Academic Career: A Brief Guide for PhD Students, Postdocs, and New Faculty, Special Publications 74, First Edition. Jeffrey J. McDonnell.

Here's what successful scientists do.

Know what science is. The late Carl Sagan often wrote, "Science is a way of thinking much more than it is a body of knowledge." What he meant is that curiosity about how things work drives most of us. Somehow, the most successful scientists find a way to enjoy that in their inner monologue. My absolute favorite definition of scientific research is "Seeing what everyone else has seen, and thinking what no one else has thought."[5] That's what it is all about. Such phrases become useful mantras for weaving science into daily observations and thinking.

Know the success equation. The author Thomas Friedman has an equation for this: "CQ + PQ > IQ," which means that the curiosity quotient and the passion quotient together exceed the intelligence quotient. In observing the most successful scientists in my field, I've found that passion and curiosity live next door to drive, curiosity, and tenacity. Friedman is close to what E. O. Wilson's "Letters to a Young Scientist"[6] will tell you, that success in science is not about genius but is rather about common sense and an inquiring mind.

Have an indomitable spirt. Indomitable means "impossible to subdue or defeat."[7] In William Ian Beardmore Beveridge's 1950 classic, *The Art of Scientific Investigation*, he discussed traits of successful scientists and notes that "a spirit of indomitable perseverance has characterized nearly all successful scientists."[8]

Work on important problems. Richard Hamming's classic essay, "You and Your Research," provides one of the most useful statements on research that I've read: "If you do not work on an important problem, it's unlikely you'll do important work."[9] That could be the one tattoo I would encourage my new graduate students to get: the importance of such a statement will never dull, never fade, never go out of style! The motto of the Helmholtz Association of the German Research Centers, Germany's largest scientific organization, with an annual budget of more than €4 billion, is "Solving the Grand Challenges of Science, Society and Industry." That effectively means "Research for Grand Challenges," not "Research for Narrow, Obscure Questions That Only I Am Interested In."

Take time off. The most successful people in academia I know take time off. Almost 100 years ago, people such as J. P. Morgan Sr., a famous American financier, would say about the need for rest: "I can do a full year's work in 9 months, but not in 12 months" (perhaps especially apt for U.S. faculty on 9-month appointments). Other early writers commenting on taking time off from research were wont to say "Freshness and originality may be lost if the scientist works unremittingly for too long … Most of us require recreation and variety in interests to avoid becoming dull, stodgy and mentally constipated."[10] Today, we might say "Work hard, play hard." Trite, but so true. The best scientists I know pursue interests outside work: sailing, adventure motorcycling, playing ice hockey, performing or listening to music, and pursuing photography. Remember the old proverb: all work and no play makes Jack a dull boy and Jill a dull girl!

Are ambitious, but not excessively so. In his classic "Advice to a Young Scientist," Peter Medawar considers ambition a "motivating force to get things done—not necessarily a deadly sin, but excess of ambition can certainly be a disfigurement."[11] Knowing that line and not crossing it is key.

Are lucky (and appreciative). Let's face it, there is luck in all of this. Some might say that we make our own luck. I prefer what Robert Frank says in his recent book, *Success and Luck*, where he notes that "lots of talent and very hard work still needs an element of luck."[12] We often don't notice the subtle chance events that allow us to succeed. Acknowledging luck bathes you in appreciation. It helps avoid hubris, a truly deadly sin in science.

I'll end with a quote from one of my favorite writers, Fareed Zakaria. He summarizes it nicely: "If you want to succeed in life, most often you need to put in the hours, develop good habits, work well with others and get lucky." That's true whether you study English, physics, history, engineering, or business.[13]

Progress = pain + reflection.

Ray Dalio, American hedge fund manager

3

AVOIDING A PHD IN ANXIETY

PhD programs are stressful. PhD students have crises. I remember as a PhD student not knowing how to turn my mind off at night, my sleep shattered as I ruminated on my work and experiments. I was working too much and not powering down my brain with other activities to achieve the peace of mind I required for sleep. After some months, I realized that I needed weekends away in the New Zealand wilderness to recharge and give myself a break. A recent survey of 3659 PhD students in the journal *Research Policy* showed that more than 50% of them struggled with their mental health during their PhD program.[14] Such crises, from mild to major, are commonplace. What is not common is discussing it. In many ways, knowing that an upset or two will happen during a PhD program can help the student know how to deal with those crises before they happen. Know

Navigating an Academic Career: A Brief Guide for PhD Students, Postdocs, and New Faculty, Special Publications 74, First Edition. Jeffrey J. McDonnell.

that when rough seas come, calm water awaits and that emotional crests follow emotional troughs.

Myriad reasons lead to PhD crises: difficulty writing a chapter; a challenging student colleague or committee member; a thorny relationship with your supervisor; stress linked to the uncertainty of what comes next. In many ways, your PhD program includes two separate journeys. One is in the subfield of your thesis, the other a personal journey, of self-discovery; of learning what makes you tick and the limits of your mental and emotional abilities as you are pushed like you've never been before.

Here are some stress-relieving hacks for your academic journey:

Know that perfection is the enemy of good. This is an oft-quoted saying but is one that applies especially to PhD students as they develop their proposal or first paper. Slaying the perfectionist tendency early on is key to handing in your draft to your advisor without fear of judgment and to keeping the process moving along at a fast tempo. Commit this adage to memory: "A good dissertation is a done dissertation. A great dissertation is a published dissertation. A perfect dissertation is neither."

Don't be a busy fool. Effective PhD students balance focus with hard work. Your acceptance into your PhD program has already proven that you can work hard. The secret now is to work smart. Many a student has lived in a constant busy-stressed state, which prevents them from making headway in their work. Solutions for this abound, but the best and simplest one, I've found, is keeping a daily to-do list. What separates a list from an action plan, however, is a time estimation for each task and a ranking of priority of each item to tackle each day. Together, these become potent ways to swallow the PhD elephant: one bite at a time.

Read. Read, skim read, speed read, and deep read. Read. Did I say read? Reading is crucial. Notes Beveridge: "The World List of Scientific Periodicals indexes more than 50,000 periodicals. A simple calculation shows that this is equivalent to 2 million articles a year, or 40,000 a week, which reveals the utter impossibility of keeping abreast of more than a small fraction of the literature which is most pertinent to one's interest."[15] He said that in 1950. Think about that. Reading is not a new challenge. And while the exponential increase in the number of journals and papers since 1950 has only worsened this situation, the Internet places all these works at our fingertips. How to read? Well, Francis Bacon had that one down in the 1700s when he said, "Read not to contradict and refute, nor to believe and take for granted … but to weigh and consider."[16]

Did I say read?

Follow the "Do-Something" principle. Most PhD students think that writing or creative output comes from inspiration and that inspiration leads to motivation, which then leads to some desirable action (i.e., writing a chapter of their thesis). Mark Manson's "Do-Something" principle reverses this theory. He argues that action (any kind of action) can generate inspiration, which can lead to motivation.[17] I've had countless personal examples of this, including a recent one in which I was flying home from China. I'd planned to work on a paper. But when I boarded, a drink came and then a meal and then the movie selection was so darn good. I'd promised myself to get that paper done but I'd lost my motivation. I then opened my laptop to work on a slide for my next talk; that action ultimately led to inspiration to work on the paper. The "Do-Something" principle had kicked in.

Know that to have a good idea, you need a lot of ideas. Keep in mind this quote (attributed to Einstein): "Creativity is more important than knowledge." But where does that creativity strike, you ask? Everywhere. Daily. Get into the habit of being open to ideas that float into your head, usually on a walk, on a bike ride, at the gym, or on a long car ride. These are gems to be mined. Reading for pleasure can also help trigger ideas and help you to consider links among and between them. This is what Darwin talks about in his memoir: "originality often consists of linking up ideas whose connection was not previously suspected"[18].

Those are my suggestions for easing your academic journey. Here are some stress-relieving hacks for your personal journey.

Proceed as if success is inevitable. Such fridge-magnet statements can help shield you from negative thoughts. Visualize, with specificity, your thesis, and your relationships, your life. (Remember, however, that my PhD is in hydrology and not in psychology: these words are experiential and are not clinically based!)

Have a fallback plan. A contingency or backup plan, or Plan B, can be comforting and helpful. For me, becoming a policeman, like my father had been, was my Plan B. It bathed me with a sense of comfort in the knowledge that if all else failed, if it all came down, then I still had a plan.

Find ways to turn it off. When I was a student, the pressures were acute at times. It was tough to turn off my thoughts at night and tough to not let my research invade my every waking hour. There was pressure around completing my PhD program in time to take on an offered job. I know that stresses for today's PhD students are even greater: the ability to stay digitally connected 24/7 keeps the burner always on high. Go to the gym or find ways to relieve stress that are right for you. One trick that has worked miracles for me is completing a detailed "to-do list" before bed. That gets it out of my head and onto paper. This uncomplicated nightly act frees my mind from having to think about tasks overnight since I know that they are safely parked and waiting for me in the morning. It's a simple strategy, but oh, so helpful.

Bathe yourself in the wonders of what you are doing. Knowing and appreciating how lucky you are to be doing what you are doing is a way to turn stress around. See Chapter 2 on the "Traits of successful scientists." Your inner monologue is a useful telltale sign of your mental state. Examine that monologue and change it as necessary to return to a thankful message to yourself and not an "oh no!".

Embrace strong values. Ethics are a key component of what universities seek in candidates for assistant professor positions (see Chapter 8). You might keep in mind a quote from Ralph Waldo Emerson: "Sow a thought and you reap an action; sow an act and you reap a habit; sow a habit and you reap a character; sow a character and you reap a destiny."[19] That is a very good one. My mother, who was a bit more to the point (and Biblical), always said, "The road to hell is paved with good intentions." This means, don't slack off and don't procrastinate. Whatever your values—adventure, achievement, and giving back—use them to help keep you grounded through your PhD journey.

Learn how to handle criticism. This one is hard. I remember craving critical feedback on a chapter in my thesis from a noted local soil physicist. It came back to me with many red marks (in pen; this was pre-track changes), and I had the strange sensation of both hating and appreciating him for them. It took me my whole PhD period to get to grips with criticism. It is the bedrock on which science is built. You must learn how to handle it, embrace it, appreciate it, and then, once you receive it, to reflect on it. Indeed, reflection is crucial for progress as a student and as a scientist, as Ray Dalio notes in his book "Principles." In fact, he has an equation for it: Progress = pain + reflection.[20]

4

GETTING YOUR FIRST ACADEMIC JOB

Who gets hired? To the applicant, this seems a simple question, with the answer being "Whoever is best qualified!" But, alas, that is seldom the case. True, the CV is what gets you to the interview stage, but after that, all bets are off for anything that's written. It then comes down to personality and interpersonal skills.

I recall the dean at the first university where I interviewed asking me, "So Jeff, who's your hero?" I'd just gotten off the plane from New Zealand and, in my jet-lagged state, spouted off the only name I could think of, Wayne Gretzky (Canada's top hockey player of all time). It got only worse when his follow-up question was, "What book are you reading?" I'd been on an Eastern mysticism kick at the time and was hoovering up books by Shirley MacLaine. I was certain afterwards that my answers had taken me out of the running. By some miracle

Navigating an Academic Career: A Brief Guide for PhD Students, Postdocs, and New Faculty, Special Publications 74, First Edition. Jeffrey J. McDonnell.

they did not, but when I was talking later to other faculty members who had been hired around the same time as me, they laughed at my answers; most of theirs had been some inspirational past president or a more sophisticated academic book!

Interviews are strange. And department members may not know what they want from an incoming assistant professor, which can provide a huge advantage to the candidate who has the ability to connect with the department and hiring committee. I've sat on many a hiring committee in which, in the end, we hired someone without knowing what they would do in terms of teaching and when they arrived, we were simply spellbound with them and their research vision. They had charmed and wowed us. Behind the scenes, a typical university department search is often rife with debate (even arguments) about the position description and requirements. One faction of the department wants one set of things from that position while another faction wants another.

Of course, there is more to the hiring process. Here's what you need to know.

The job ad. Read between the lines. Identify what is essential and what is desirable. The shortlist of candidates often hangs on the balance of these "what is required" words, where the ones who make it to the shortlist tick all the mandated boxes. In your job letter, parrot back language in the ad and explain how you are *the* person for the job, based on your background and qualifications. Writing a job letter is analogous to writing a research proposal. If you do not hit what is asked for, it will go into the bin.

Your CV. It must be perfect and free of typos. You cannot imagine how many CVs I have examined while on search committees only to find typos, inconsistent referencing, and other disqualifying errors. Any imperfection in your CV (or job letter) will paint you as someone who does sloppy science. Have a colleague or friend with a sharp eye proofread *everything* you submit with your application.

The phone interview. You must be enthusiastic and articulate. You must be seen as sharp and bright. The key is to perfect your humble brag: be self-effacing, but present crucial information that touts your accomplishments and potential. Your research on your would-be department will help you shine brightly here. Knowing who those department members are and how you fit into their culture will help enormously for getting you to the next level.

Your elevator pitch. The phone interview and the mini-encounters you have with faculty members at the interview will necessitate a condensed statement of your research, known as an elevator pitch. Many times, these are disastrous. The candidate drones on, saying that she/he does this, *and* this, *and* this, and by the third "*and*," the listeners' eyes are glazing over. Randy Olson has a solution for this. He calls it the ABT method: And, But, Therefore.[21] Find out more about this technique in Chapter 12.

Research the department. Your knowledge of the department is directly proportional to your ability to wow both individuals and the collective when you visit the campus. Get familiar with details about the faculty members whom you may meet with one-on-one, and especially with members of the search committee, such as where they earned their PhDs, so that you can use that information during small talk. Know what they study so that you can ask them about themselves and their work. You must appear as an interested equal.

The seminar. The best interview seminars teach the audience. Why? Because many in that audience will not know your subfield. This is not a conference talk, so leave technical jargon and specific details out. The best seminars bring everyone in the audience along with the speaker to the end of the journey. These seminars tell a story. They also have visually captivating slides and are exceptionally well rehearsed and well paced (a slide per minute optimal).

The teaching example. This must shine. Pretend that faculty members in the audience are students. Where a research talk is paced at one slide per minute, this must be at one slide per 3 min (or longer). Aim to conveying one thing through your lecture: one piece of crucial information that the faux students can take away from the class (and discuss later in their committee deliberations).

The interpersonal interactions during the interview. Know that most people you'll speak with during the interview process are likely to be thinking, "What can this woman/man do for me?" They'll ask themselves, "Can I work with this person? Will they be a person to serve on committees with? Will they be a good departmental citizen?" During your one-on-one interactions, you must convey, often through small talk and showing interest, that you can tick these boxes and that you have value, to the organization and to them as individuals.

The offer and negotiations. My first job offer came in a letter. My only negotiating tactic was to immediately send a reply stating, "Yes, I accept!" I did not counter any part of the offer, I simply accepted all of it. This was a poor strategy. At time of offer, you, the chosen candidate, hold the power.

Conduct negotiations. Do this through the lens of how your requests will let you hit the ground running once you arrive at the new university. So you'll need to pitch your start-up package and salary requests as seed money for the fame and riches you will bring them in the future.

Research salary information online. Many states, provinces, and countries post it, and data is king. Know your comparators so that you can speak to those numbers in any counteroffers you wish to pose. Know too that your start date on campus can often be delayed to suit your postdoc or other needs. They want you. They will often wait.

Negotiation particulars. Much of all this depends on the country in which you are working or applying to work. Most of what I state above is relevant to the U.S. job market. In Canada and Australia, start-up packages are somewhat less negotiable. In some European counties, there is often no start-up package. In the United Kingdom, where interviewees are usually grouped together on campus on the same day, there are fewer formal social interactions. And in some countries, the salary scale may be locked in so that salary negotiations are not possible.

Regardless of the specifics, you need to treat the process of getting your first academic job *as a job* and perhaps as the most important research project of your career. Act accordingly. Know that personality will win the day at the interview stage. If I had one phrase to keep in mind, one that my colleague likes to use is, "You must sparkle." Plain as that. Oh, and choose your heroes and favorite writers carefully.

5

THE ACADEMIC MATTHEW EFFECT

Everyone has heard of the Matthew Effect, the well-known axiom, "The rich get richer; the poor get poorer." Research in the 1960s by Robert Merton on scientific communities showed that renowned scientists and institutions tend to attract "inordinate attention and resources, leading to further accumulation of prestige, which in turn attracts further resources."[22]

But early career scientists get little information about how this plays out for them, especially in the first few years between PhD, postdoc, and assistant professor. In effect, it comes down to the fact that "initial advantage tends to beget further advantage,"[23] as many sociological writings discuss. What does this mean? It means that exploding out of the starting gate is crucial. And those who do start strong have a cumulative advantage that accelerates with each passing year. It is important to define your role, quickly, and to not allow someone else to define it for you.

Navigating an Academic Career: A Brief Guide for PhD Students, Postdocs, and New Faculty, Special Publications 74, First Edition. Jeffrey J. McDonnell.

If research is your interest and focus, you want to be seen straight away as a research superstar. The sooner you can get grants, get papers out the door, get students on board, and get the students succeeding, the better, as your department's perception of you early on can define your teaching load for a lifetime. You need to separate yourself from your colleagues, because the effects are multiplicative.

My small Matthew Effect. I was unaware at the time, but in retrospect I can see the Matthew Effect playing a role in my own academic life. I got lucky with a string of research grants in the first two years of my assistant professor appointment. All of a sudden I had more than $1 million in funding (a lot back then), and, given the importance of indirect costs in fueling the U.S. academic enterprise, I was seen as a rain maker, someone who could generate significant income for the university. As a consequence, my teaching load lightened and, with less teaching, I had more time for grant writing. With more time for grant writing, I got more grants and, as a result, wrote more papers. You see how this goes. (Chapter 9 does describe the downside of being awash with research funding. This is also where I learned the inverse relation between grant size and impactful science outputs.)

In my experience observing others, I've seen that the first few years of one's career disproportionately affect what comes next, in one way or another. If you are free to focus and to work hard in the first few years, then the compounding effect of paper citations (similar to the concept of compound interest) helps drive your standing in future years. But, as in the case of retirement planning, if you wait until mid-career to save or publish, you will need to contribute much more at that point to overcome the absence of compounding effects that would have accrued to earlier savings and publishing.

Networking and the Matthew Effect. There is a link between networking and the rich getting richer. The more quickly one gets connected, the better. I remember my first big conference in Europe. I knew no one. I went to the sessions alone. I ate alone. I'd come from a PhD experience where I was my advisor's only PhD student, and he worked in a completely different field. I had no idea how to network.

My next meeting was a small specialty conference of about 100 people. What a difference! Organizers had structured the meeting as if everyone was marooned on an island and had to eat together and stay in the same hotel. My advice is to find those small focused meetings (ideally on your thesis topic) early on. I met five of the most important collaborators I have worked with in my career at this small specialty conference, all within a year of completing my PhD. I was also able to meet all the big names in my field there too, and in the end, I knew them and they knew me before the five days were up. It was transformational.

So why is networking so important for a fast start? It's your gateway to collaboration. There are many good books on networking, such as Alaina Levine's "Networking for Nerds."[24] But my take-home message for early-career networking is find people to write papers and proposals with. For me, my best collaborators have been people I enjoy spending time with, even if we don't talk science. Collaborators become friends.

Networks and the Matthew Effect also explain why some research groups become such dominant forces. Junior scientists develop by leaning on their network and, later, by turning out students and postdocs of their own who also have strong networks and are primed for early career success. "If you come from a non-networked lab, you must find ways to get networked because such connectivity helps differentiate even when people are exactly equal in talent and drive."[25]

So, whichever of the 26,000 or so universities worldwide you now find yourself in (fun fact: there were only about 500 universities in the world in 1945), go for it. And, as my fellow Canadian and the world's greatest ice hockey player ever, Wayne Gretzky, notes, "You miss 100% of the shots you don't take."

The world has problems; universities have departments.

*Gary Brewer, Professor Emeritus,
Yale University*

Part II: Surviving and Thriving in the First Few Years

A scientist's life would be a happy one if [she] he had only to observe and never to write.

Charles Darwin, 1809–1882,
English naturalist

6

PAPER WRITING GONE HOLLYWOOD

"So you want to be a writer?" one of my professors asked me when he learned I was interested in a career as an academic scientist, a pointed warning that a life of science is also a life of writing. But even knowing this in advance, I found that writing was a challenge as I made my way down the tenure track. I had trouble finding stories in my data sets. Even when I had a good tale, I struggled to tell it. I tried starting with the opening sentences and hoping I'd make it to the paper's end. But more often than not, I wrote my way down many blind alleys. My permanently unfinished papers outnumbered my published ones. Worst of all, I was not helping my PhD students and postdocs learn the craft of proper writing.

My big break came shortly after getting tenure. In a passing conversation, a senior colleague mentioned that his process for writing research papers centered

Navigating an Academic Career: A Brief Guide for PhD Students, Postdocs, and New Faculty, Special Publications 74, First Edition. Jeffrey J. McDonnell.

on structure. Rather than focus on words and sentences, the part of writing that so bogged me down, he highlighted the importance of outlining the overall story to be told. I had thought that the standard paper structure, introduction, methods, results, discussion, and conclusions, was enough to keep me on track. But my colleague helped me realize that, even with those sections, there is still enough freedom to get stuck in writing cul-de-sacs.

I now see each of the standard paper sections as its own Russian nesting doll. Writing papers is easiest when you spend considerable thought and time stacking all these pieces first. I call it the top–down writing approach.

Each of my group's papers now starts with a storyboard session at a whiteboard. I pretend to be a big-time Hollywood producer and ask the PhD student or postdoc to play the role of would-be movie director pitching a new movie to me. Their pitch must answer three questions: What is the status quo? What is wrong with the status quo? How does this new paper go beyond the status quo?

This approach helps frame the story and place key figures and technical findings in context. Balancing each of the status quo elements is a great way to set up the introduction, often the toughest section for early-career scientists to write, and to lead the reader to the research questions or hypotheses. Say too little about what we already know and we risk losing a large audience who may be unfamiliar with the topic. Too little about what's wrong with the current state of knowledge and the reader may wonder why we need yet another paper on that topic. Too little about how the work goes beyond what others have done and the novelty is unclear. The result is a roadmap of the novel elements in the work, which brings the discussion, the other tough section for the writing newcomer, into final focus.

Once the pitch makes sense, we go back and forth stacking the Russian dolls on the whiteboard until the outline subheadings become paragraph topics, with every paragraph explicitly represented in the outline. Honing this outline prior to any writing allows us to determine whether the research story resonates from start to finish. We might spend days or weeks on the outline to get it right, but it's time well spent. The slavish adherence to nested headings shows at a glance whether the paper makes a clear and worthy contribution; whether the title, objectives, and results are properly aligned; what figures are truly essential to the storyline; and whether the message hums. Writing then becomes a much easier process of filling in the blanks. The paper is effectively finished before the sentence writing starts.

I haven't mastered the writing game, and I am still constantly learning. But the top–down approach has been a game changer in my group. Now, when a new grad student indicates an interest in an academic career, I ask, "So you want to be a Hollywood producer?"

7

THE PERFECT DEPARTMENT

The world has problems; universities have departments. So goes the old saying. Many early-career scientists coming into their first department imagine that, like first love, it will sustain them. Sometimes it does; many times it does not. Turf wars, the intellectual ballistics of faculty meetings, old rivalries, and non-collegial behavior within the unit can sometimes mar one's journey down the career track.

While some departmental situations will be better than others, most new faculty are confronted with *some* issues in early career. Certainly, as I entered my first university department, I naively expected it would be bliss, with all my colleagues appreciating what I did in research and teaching and all of them getting along.

Now, with some gray hair and perspective, I can see how unaware I was of the human side of department life. When I send off my PhDs and postdocs to faculty positions, I do so with the realization that their worlds are about to get much

Navigating an Academic Career: A Brief Guide for PhD Students, Postdocs, and New Faculty, Special Publications 74, First Edition. Jeffrey J. McDonnell.

more complex, with increases in the flow of information and an increase in the number of delicate interpersonal waters to navigate.

We do not teach these things in graduate school. Of course, we can recommend that our PhD students and postdocs read the old Dale Carnegie book titled "How to Win Friends and Influence People."[26] But I think we can do more in terms of having fulsome conversations about these topics.

So, with the notion that soft skills are essential for surviving in departments in the hard sciences (and all departments, for that matter), here are a few thoughts that I pass on to my students who are off to their first faculty job.

Be tirelessly optimistic and practice extreme kindness and extreme empathy. Show an interest in your colleagues and in what they do. Be generous with compliments when applicable.

Say hello to others in the hallways even if they do not lift their heads. Persist in this regard; whether or not they acknowledge you, acknowledging them makes a difference.

Smile as much as you can. It is remarkable how this alone can ease bumps along the way.

Stay neutral. Old rivalries that predate your hire will sometimes exist. Being blissfully unaware of such things and floating through departmental gatherings oblivious to such matters can be very effective.

Avoid participation in any talk behind another colleague's back. Doing so is the road to ruin.

Form strong networks within your department. This is usually with other early-career faculty and recent hires. It can be helpful in and outside the work environment.

Treat each colleague around you as if they have something to teach you. This helps maintain humbleness to learning and can disarm any power hierarchies.

Being known as a positive, uplifting, and enthusiastic force in the department is what you are striving for. Academic life unfolds much more easily in such a light-hearted mode. As simple and perhaps self-evident as these suggestions may be, they are the ones of most value in our daily mantra.

But what if your kindness and charm offensive does not work and, like me over the years, you find that in some instances, collegiality in your department is as thin as the soup made from a starving pigeon?! Or maybe someone is rolling their eyes and sighing whenever you speak at a faculty meeting? Or a colleague is unusually abusive to one of your graduate students? Or the department perspective on research is locally focused and conflicts with your more progressive or global view?

There are no easy solutions to these problems. But there are a few tricks I have learned along the way, where in spite of some issues (and all departments will have some issues) an imperfect department can still be perfect for you.

Fostering your relationship with your department head is critical in helping you navigate fraught waters. They can moderate a discussion with a problematic colleague or can facilitate a crucial conversation.

Finding a way to align your work to the university's strategic plan. This can make you bulletproof to departmental oddities and sometimes parochial outlooks.

Fostering networks and engaging key people across campus that can be your saviors if the department is problematic. It can also help you better learn the university culture.

Look for reasons behind strange actions. This can help explain why someone might not like you: you may have been hired at a salary rate higher than a senior faculty (due to salary compression issues at your university); your research area may be taking off but others could see that acceleration was at the expense of the support of their research area; you have won some recent award and colleagues are envious of the honors, which can sometimes be viewed by others as a zero-sum game. Be empathetic about why colleagues may be exhibiting certain behaviors. Work the problem.

If you find yourself in a difficult department, remember the useful phrase: "Illegitimi non carborundum" ("don't let the bastards grind you down"). You can always leave. And it is not the end of the world if you do so. But before you do, know that like the Loch Ness Monster, no perfect department exists.

Great people have great values and great ethics.

Jeffrey Gitomer, American author and business trainer

8

ETHICS IN THE REAL WORLD

When I was in my late 30s, I was a professor at a SUNY (State University of New York) campus. I was awash with research dollars. Much of my funding was for applied work in the New York City watershed. I had excess summer funding, to the point where I could bankroll some of it to use in a "slush fund" account to fund student travel, buy lab equipment and the like. I had a summer cottage on the

Navigating an Academic Career: A Brief Guide for PhD Students, Postdocs, and New Faculty, Special Publications 74, First Edition. Jeffrey J. McDonnell.

St. Lawrence River where I was wondering about how groundwater might bubble up from beneath the river and affect some of the water mixing models I was using.

This related to my basic research interests and other lines of U.S. National Science Foundation-funded research. One day I had the bright idea that I could use some of the slush fund money to buy SCUBA equipment and to sample the bubbling groundwater on my time off. I thought nothing of the optics of this with my engineering colleagues, many of whom did all their work in front of computers with no field work. But I later learned that my buying SCUBA equipment did not sit well with some of them. And only a few years after I had left the university to join another, did I realize the impact this had locally, for some, on my reputation.

The American Geophysical Union (AGU) and other scientific societies have recently been promoting policies on scientific ethics. I served as President of AGU's Hydrology Section and, in discussing these with other Section Officers, we quickly recognized that in our research, educational, and community roles, ethical issues have often fallen into gray areas. What to do was often not obvious, and a good case could be made for different options. Looking back, we presumably like most other senior scientists, handled some issues well and others poorly. Hence, we suggest that discussing scientific ethics by generalizing past dilemmas can be an effective way to engage students, postdocs, and early-career faculty members in such discussions. Of course, there is the venerable "On Being a Scientist" book by the U.S. National Academies[27] that does this with complex examples.

Below are some of our (AGU members') personal experiences that we hope might stimulate discussion and debate and might be a way for more groups to begin discussions about scientific ethics in their professional societies, student clubs, and departmental coffee rooms.

You have identified an interesting phenomenon in publicly available data and developed a good explanation. Before you can write this up for publication, you get someone else's paper to review. The paper notes the same phenomenon, but offers an explanation that is clearly wrong. Pointing out the error will almost certainly lead the other authors to your answer. What do you do?

You are finishing a paper just in time to meet a journal's special-issue deadline. A coauthor suggests doing further analysis that is a direct extension of what you've done. To your surprise, the additional analysis gives a different result. You're pretty sure there's a problem with the new analysis, but can't figure out what's wrong. What do you do?

You are presenting a poster at a meeting showing your data and offering possible explanations, all of which you know are inadequate. Someone looking at the poster offers a simple and much better explanation. What do you do?

You are putting together the authors list for a paper you're finishing. One of your fellow students gave you some of the data, another helped with the data analysis, and the third made several good suggestions. What do you do?

You are a student discussing with your advisor where to submit a paper. You are tempted by a prestigious open-access journal. Your advisor is amenable, but explains that paying the fees would spend the funds that would otherwise fund your trip to an upcoming conference. What do you do?

You are a faculty member chatting with a graduate student, who complains that they gave a paper draft to their advisor several months ago and has yet to get comments back. The topic is far from your expertise, and the advisor is a friend. You would like to help the student but don't want to cause them difficulties. What do you do?

You are a faculty member who encouraged a younger colleague to apply for a position in your department. To your surprise, their thesis advisor wrote a very negative recommendation letter, which takes your friend's application out of consideration. You want to tell your friend about their advisor's letter. What do you do?

A journal has rejected a number of papers you have submitted. They then ask you to review a paper for them. What do you do?

You are a faculty member talking to an undergraduate who seeks advice because of their mediocre grades in another department's lab class. Other students have learned that getting an "A" requires fudging results to reach the expected accuracy. However, the student is reporting the results as obtained, causing lower grades. What do you suggest?

You are a faculty member who has identified several students cheating in class. Under university rules, you are not allowed to take any action except reporting them to the dean's office. Your experience has been that the dean's office rarely imposes penalties even in such blatant cases. What do you do?

Discussing such questions leads rapidly to a list of the benefits and costs of different approaches to the issue at hand. This in turn leads to the challenge of how to choose between approaches, given that ideal and practical approaches often differ. In many cases, how the different approaches would actually work is unclear. Ultimately, there are no unique or right answers. Perhaps the best criterion is to ask, "How would I feel if 'this' was on the front page of my local newspaper?"

My volunteer leadership colleagues and I at the AGU encourage scientists and scientific societies to discuss the spectrum of ethical challenges with their mentees, through special conference panels and sessions, in departmental seminars, and around the coffee table. There is much to be learned through personal examples and honest reflection on choices made and what one might have done differently in hindsight. Discussing scientific ethics in the real world may help engage our community in raising awareness and devising proactive solutions to common dilemmas.

Proceed as if success is inevitable.

Anonymous

9

RESEARCH BRAND IDENTITY

Navigating an Academic Career: A Brief Guide for PhD Students, Postdocs, and New Faculty, Special Publications 74, First Edition. Jeffrey J. McDonnell.

In my first few years slogging away on the tenure track, I flitted from one research question to another. In an undisciplined pursuit of funding, I wasted my energies, and, sadly, papers, on unrelated topics. I began to understand how a successful research program worked only after I became an associate professor at another university. I realized from observing others that it was critical to find one's focus and voice and have it heard, quickly. Now, when my own students and postdocs ask me how they can stand out among the many young scientists jockeying for position, I tell them to learn from my early struggles, and to define their research brand identity.

A research brand identity, in the best scientific sense, is your central mission: the particular branch of research you'd like to be truly excellent in and known for. Equal parts research focus and career vision, a brand identity is a map of a few key subareas where you strive to make a difference over a sustained period. By crafting a research brand, you make it clear to others how they should define you, and you help them remember who you are and what you do.

As a young hydrologist, I worked on topics as varied as soil erosion, snow physics, and the remote sensing of water uptake by plants. When I finally found my focus, I concentrated on three simple questions: Where does water go when it rains? What pathway does it take to the stream? How old is the water in our streams? At first these questions were narrow; I explored just beyond the boundaries of my thesis work, in a space where I had established some credibility. Over time the questions broadened. I now explore where plants get their water and the role of geology in defining water storage and release. These core issues have sustained me over my career.

The trick is to find the optimal research "brand width": too broad and you blend in with others and lack a niche to carve out; too narrow and you seem too limited. You want some connection to your advisor's research brand, but you also want to move in a new direction that honors your unique skills. The old adage, "Only do what *only you* can do," can be helpful as you ponder this.

After you have identified your brand identity, getting it out into the world is relatively straightforward: get good results, publish papers, and give talks that all map to your theme. But there is much more. The first step is describing your brand succinctly on your research homepage. The second, and more active, step is promoting it through a set of activities directed at your peers. You might write a journal commentary, perspective paper, or a review linked to your brand. You might guest edit a journal's special issue or edit a research text. Or you could organize a conference session or a boutique research conference focused on your theme. The idea is not to do just one of these things but to do as many as you can.

All of these activities build a whole that is greater than the sum of the parts. Come tenure time, you want to be known for something, and you want external reviewers examining your dossier to already know of you. This strategy will almost guarantee that.

Finally, a research brand identity remains helpful as you advance. As more opportunities come your way, your brand can help you filter them, keeping your main thing the main thing. The brand helps you determine what to say no to, a skill that becomes much more important as you grow more sought after over time. And just as a research brand identity is a way to describe your ambitions when you start out, it is also a way to sum up your body of work later in your career.

What is important is seldom urgent and what is urgent is seldom important.

Dwight Eisenhower, 1890–1969,
34th U.S. President

10

THE 1-HOUR WORKDAY

When I was a new assistant professor, I felt constantly overwhelmed. I had classes to teach, relationships with new colleagues to navigate, a lab group to assemble, and an infant at home, not to mention research to conduct and papers to publish. To get ahead, I took on any opportunities that were offered, including membership on various editorial boards and professional committees. Despite working like a madman, my productivity as measured by paper output was meager. I simply could not find time in my day for undistracted writing. And when I did find the time after an extended stretch away from writing, the warm-up period to get back into the paper was often long, further slowing my progress.

At first I thought that this kind of frustration was routine for academic researchers. But as the years passed, I noticed a few senior colleagues who published with impressive regularity and always had a paper in the works. When

Navigating an Academic Career: A Brief Guide for PhD Students, Postdocs, and New Faculty, Special Publications 74, First Edition. Jeffrey J. McDonnell.

I asked them what their secret was, I found that they prioritized doing small amounts of focused writing every day. I've since developed my own version of this approach. I call it the 1-hour workday, referring to the short, sacrosanct period when I do what I see as the "real" work of academia: writing papers.

First thing in the morning is when I'm at my mental best, and when I'm still most in control of my time, I now use the first hour of my day to write. For me, it's best done from home. I've developed something of a ritual: I wake up early, make an espresso, and write until I'm spent, or until distractions like email or the day's deadlines and meetings start to intrude. This is usually about an hour, some days a little less and some days more. I've found that, like hitting a ball in golf, regular writing is easier if I tee it up. I plan my early morning writing the night before. It is in my calendar and on my to-do list, with details about which paper and section I will be working on.

This routine has transformed my work life. Instead of the frustration that frequently plagued me early in my career, now, no matter how work proceeds after I've completed my writing time, I go home at the end of the day with the satisfaction of having accomplished something.

I have in no way mastered the writing game, but my 1-hour workday has certainly increased my academic output. And by keeping me focused and in practice, it has improved the quality of my writing and made the process much more enjoyable. It also offers an opportunity for deep thinking. I remember rarely having any such thinking time when I started out as a professor, but now I find that my daily keystrokes can lead to new ideas. When I string together days of successful writing, ideas flow and new connections present themselves even when I'm away from my keyboard, particularly on my bike ride to work or when I'm reading for pleasure.

Many days, "writing" means editing and revising the work of others. This work can sometimes be a slog, but I keep my spirits up by thinking of it as a game of table tennis. My goal is to return the serve, when done quickly, this greatly improves the game.

In any sport, one must stay toned and conditioned. If I fall out of practice, I quickly lose that fitness, and my writing and editing become labored. So, even though distractions abound, I protect that daily workout at the keyboard during the first precious work hour of the day. I've learned that writing does not need long stretches of uninterrupted time. Focus and regularity are what matter. I now advise my PhD students and postdocs who are going on to faculty positions to adopt daily writing as an early-career habit so that they don't repeat my years of writing frustration. At any career stage, a daily writing ritual can help improve performance, and, perhaps most importantly, job satisfaction.

11

ORCHESTRATING A POWERFUL RESEARCH GROUP

As I entered my assistant-professor years and worked to assemble my research team, I considered each candidate individually. I took on students based on their grades and test scores, and my relationships with them were one-on-one. I didn't consider their teamwork abilities or soft skills, or the group dynamic as a whole. This approach gave me a somewhat productive lab group as measured by single-member outputs, but over many years, I came to appreciate that the collective matters, a lot. Beyond the individual output of the graduate students and post-docs lies a parallel universe of teamwork, peer-to-peer mentoring, and, most important, discovery for the research group as a whole.

I found this quite by accident, 10 years into my faculty appointment, with the arrival of a European postdoc who insisted that the group have daily morning coffee like he had "back home." This ritual evolved from nonacademic conversations over

Navigating an Academic Career: A Brief Guide for PhD Students, Postdocs, and New Faculty, Special Publications 74, First Edition. Jeffrey J. McDonnell.

pastries to daily check-ins about what the group was working on to discussions of new ideas. Over the years, these conversations have been the most satisfying part of my job and have led to some of my group's better papers.

I now think of each group member as a crucial puzzle piece for my collective. I assemble teams of individuals with different but complementary scientific backgrounds and play off the (healthy) tension between them, where they question one another's approaches and perspectives. Research group members will have their own theses, projects, and papers, but one can orchestrate a group dynamic that promotes discussion about where the field should be headed and the best new questions to ask.

The first step toward creating this environment and fostering a powerful research group is building relationships. Regular social activities outside of work can help break down walls and create a team spirit. Weekly lab meetings, morning coffee, group lunches, or Friday after-work beers can engineer serendipity across the entire group or across smaller subgroups that head into new directions with curiosity-driven side projects. The team building also creates a sense of safety, trust, and belonging. In my lab, there are high expectations for unselfish cooperation and maintaining one another's reputations. We adhere to the old adage that if you do not have something nice to say about someone, then say nothing at all. Foibles are tolerated; disagreements are settled quickly.

Lab diversity, be it scientific or cultural, can make building community challenging, but that diversity itself can add immense power to the team. Certainly I have found that my most productive groups over the years have been the ones with the greatest gender balance and range of cultural and scientific backgrounds. A group leader can learn to leverage this diversity and draw out ideas from some who may be timid in group discussions. Something as simple as not letting anyone dominate during discussions can help. Inviting a gifted member to throw out an idea and having the group discuss it can also be effective.

Using the powerful group as a way to think is like conducting an orchestra. It involves assembling a varied group of musicians, each with solo skills, and helping them play together, creating a piece I could never accomplish myself or with a single student or postdoc.

I have in no way mastered the powerful research group, which is an evolving and ever-changing thing. New lab members bring in new opportunities and challenges. But one thing I now understand is that as old members move on, the powerful group extends well beyond the faculty member's home institution. The group becomes a ready-made network for collaboration among lab alumni who go on to develop their own orchestras, repeating the cycle of the powerful research group elsewhere.

12

PROPOSAL WRITING HACKS

So you want to be a salesperson? That is a question I might pose to a new graduate as something analogous to the questions I received myself about being a writer that I mention in Chapter 6. A life in academia is a life in sales: of your ideas through your proposals. One does not usually get this sales training in a graduate program, but winning grants is the lifeblood of research. And writing proposals is a different kettle of fish than writing journal articles.

So here are a few hacks I learned along the way:

Know the target. The proposed research for any competition must be fundable, laudable, and achievable. Let those three words sink in for a tick. Fundable means that you are asking for money within the parameters of what is possible from the granting agency. Laudable means that the research is meritorious in some way, it

Navigating an Academic Career: A Brief Guide for PhD Students, Postdocs, and New Faculty, Special Publications 74, First Edition. Jeffrey J. McDonnell.

solves a problem (if the grant would fund more applied research) or tackles an important problem in a novel way (if the grant would fund more basic research). And achievable means that you can complete the research and realize your stated goals within the constraints of funding. Many an application is axed for being unachievable, as reviewers simply do not believe that the proposal could be solved in, say, three years and with x dollars.

Read the call: It is important to know, too, that if you are responding to a specific proposal call that has programmatic objectives, then you must respond to what is sought. This seems so straightforward, but a busy early-career scientist's enthusiasm for their own research can cloud their ability to closely read and understand exactly what the call is seeking. Writing a proposal for a detailed call means reading that call over and over and submitting a proposal that responds to each and every detail, every nuance of that call. It's science in a straitjacket, but if that is the coat you are given, you must wear it. True, some proposal calls are simply looking for good ideas from the bottom up. This gives you more rope, but sometimes enough to hang yourself!

Sell your ideas through the ABT format. Randy Olson has a book, "Houston, We Have a Narrative."[28] It may be the single best resource for proposal writing (and academic "pitches" in general, as discussed in Chapter 4). Many ideas are set up and sold by saying that one will do this, *and* this, *and* this, *and* this. For example, "We will study water flowpaths on a hillslope, *and* we will use isotope tracers to track the water molecule *and* we will age-date the water in the stream."

Randy's message is one as old as time: Tell a story through the And, But, Therefore (ABT) structure, the same one used by the ancient Greeks and by today's Hollywood screenwriters. That "and, and, and" story can be transformed with ABT. For example, hydrologists have studied the water balance for 350 years. *And,* we have learned much about evaporation, rainfall and streamflow and the annual rhythm of water cycling in catchments. *But,* recent work with stable isotopes has shown that water remains in some compartments for centuries to thousands of years. *Therefore,* we propose a new water balance model that takes into account these water ages.

The dramatic ABT tension plays into the way people are wired to receive stories, just like how country-music enthusiasts are wired to hear a 1–4–5 chord progression! Just listen to any country song. If it's written in the key of C, then I'll bet you a beer that the song goes from C to F to G to C in some combination.

Nontrivial hypotheses. After serving on U.S. National Science Foundation review panels and evaluation teams, I learned one common refrain of Program Managers: they were looking for "sexy questions." Meaning, they wanted proposals with nontrivial hypotheses, ones that would shift the paradigm. How would this move the dial? We're talking here about high-risk, high-reward research. Many managers say simply that the hypotheses they see are too trivial.

How compelling and persuasive, you ask? Creating drama in a proposal is what it is all about. Writing in the active tense helps propel that sense. Important too is making clear to the reviewer what key advances will be made in the field if the proposal is funded. And what others will be able to do that cannot be done now, if the research is successful now. It is equally important for the reviewer to see what will be robbed from the field if the proposal is *not* funded. How will they as a reviewer do a disservice to the field if the work is not supported? Did I say drama?

Pilot work. In increasingly competitive proposal competitions, pilot work, small bits of work that show some hints that your ideas are possible, can be a way to sell an idea. All things being equal, if you are reviewing two high-risk proposals that are equally well written, which one would you fund?

Easy-peasy writing. The new proposal writer may want to sell their wares by rolling out all the jargon from the journal literature they are reading. Avoid this trap! The proposal must be easy to read, easy to follow. The key is to not lose the reader anywhere along the way, not in an equation, not in a long detailed paragraph, and not in a long sentence. Short. Sweet. Easy. Use the largest (not the smallest) font allowed. Make it inviting to read and digest.

And most of all, do not ever make a reviewer feel dumb in any section of the proposal. Your proposal will be reviewed by a wider audience than you would ever think possible. Teach each audience member as you sell to them. You do not want to lose them anywhere along the way, or they will get up to get a coffee, check Facebook, or open up their e-mail. They will be distracted. Keep them glued to your page like a James Patterson novel.

Words from a grand master. I worked with a grant writer named Donn Forbes[29] on two large and successful proposals, starting about 10 years ago. I'd been writing proposals for nearly 20 years by this time. Working with Donn was like apprenticing with a grand master, he's brought in close to $1 billion in funding over his career. He showed me what a lightweight I was in the proposal-writing department! He opened up new insights I did not know existed. I include some of his advanced hacks below:

- *On visualization*: Donn advocates helping the reviewer visualize people doing things. Rather than telling them what you will do, he advises to "include the names and titles of who will do what and how they will work together." Provide the ability for a reviewer to see into your lab and "view" what you are doing. He likes to quote Joseph Conrad's "Heart of Darkness,"[30] in this regard: "My task is to make you hear, to make you feel - it is, above all, *to make you see*."
- *On the Work Plan*: Donn's most important part of the proposal is the Work Plan. His advice is that "at every opportunity, point out the innovative aspects of your work. Innovation is a key criterion and if your Work Plan isn't innovative, why should the agency support it?" Any Work Plan, of course, should outline the proposal's main goal and list the objectives for reaching that main goal. But Donn advocates going one step further: "Frame each objective as a discrete step with a quantifiable definition of success." He advocates

"beginning each objective with a verb. Then follow with a statement of who will complete it and when. Then, after listing the objectives, explain in exquisite detail how you will accomplish each one." And finally, divide the Work Plan into phases to make it easier to understand.

- *On the psychology of review*: If you can write the proposal simply and clearly enough, the reviewer will actually read it and allow you to communicate your vision and your technical plans. But Donn's take is that "the reviewer too will experience satisfaction from understanding your ideas. And the reviewer will have the opportunity to think about your proposal within her or his own frame of reference and to generate her or his own ideas about why the proposed work would be valuable. The reviewer wants *to enhance her or his professional status* by funding an excellent proposal. You want to give the reviewer the opportunity to be a hero!"
- *Lastly, and perhaps most importantly from Donn's advanced hacks, you want to "recite the journalist's mantra*: explain who, what, where, when, why and how."

Are you shovel ready? I have an outstanding Vice President of Research at my university whose mantra for proposals, especially big ones, is to be "shovel ready." Like a construction project where much of the planning prework is done, it means that grant writing can begin quickly. You are ready to pounce. Writing grants is about being shovel ready with your ideas, with your proposal basics, with your collaborators, and with your networks.

Now, who wants to buy a shovel?!

Part III: Tenure and Beyond

Leaders don't create followers, they create more leaders.

Tom Peters, American business management author

13

MENTORING THE MIDDLE

I felt like my mid-career progress as a hydrologist at the State University of New York in Syracuse was on track in the mid-1990s. I'd received tenure four years after finishing my PhD.

So when I applied for a U.S. National Science Foundation mid-career grant that required a senior colleague in my field to review my CV for inclusion in the proposal, I reached out to a luminary who was a member of the U.S. National Academy of Sciences. I was excited about what I thought was a strong package, and about what I fully expected would be a strong endorsement.

I don't now remember how I came to see the form he faxed back, but his response is still indelibly stamped on my mind: he had ticked a grudging "good" on the CV evaluation box, next to a scratched-out "poor." I had been certain that I would get an "excellent."

Navigating an Academic Career: A Brief Guide for PhD Students, Postdocs, and New Faculty, Special Publications 74, First Edition. Jeffrey J. McDonnell.

Aghast, I wondered what had happened. My transition from assistant to associate professor had gone swimmingly and even ahead of schedule. What had gone wrong?

Years later, I realized two things: the standards of excellence in my department were not the same as standards elsewhere. My institution's bar was lower. Looking back, I see now that my senior colleague's department had faculty members who were my age and at my career stage who were publishing regularly in high-impact journals, while all of my work was in disciplinary journals.

And, perhaps more importantly, the reviewer had determined that I had been doing largely more of the same, grants in, papers out, since getting tenure and hadn't taken any steps toward leadership in my field. If I read between the lines, he may well have thought that I was not going after the big, important questions in my field. Admittedly, I was letting grant opportunities drive what I worked on.

But unlike in early-career stages, when mentors and mentoring pointers and targets abound, mentoring of the mid-career scientist in academic or government research is largely absent. My own mid-career disappointment (I did not get the grant) showed me that beyond my local institutional bubble, new things were expected. But what?

Now, when I mentor mid-career colleagues, I tell them about my experience and try to help them to understand the differences between early-career expectations, which tend to center on numbers of papers, numbers of grants, numbers of students supervised for earning tenure, and the new set of leadership expectations that define mid-career advancement.

But mid-career can be a tough transition: gone are the days of phased-in teaching of the first few years. The weight of committee work, teaching, research, peer review, service to professional societies, and more can begin to take its toll. On top of that, there is the mid-career churn: the money-in-papers-out treadmill that can lead to tenure but that may prevent you from thoughtfully reflecting on the big questions in your field. Many get trapped into what they let the job become rather than reshaping it into a new path forward. Like me in my earlier days, they are not deliberately controlling what they are working on. Malaise can set in.

Here are a few ideas for ascending to the next level.

Shift to a leading role. Think about where your field should be headed if you've not done so already. Ask yourself, "What are the grand challenges in my field?" "What don't we know; but need to?" and "How can my work attack them?" Instead of focusing on the number of papers you publish, you will need to concentrate on the impact of your published works as they relate to these grand challenges.

You will also need to catalyze the international scientific community around these questions. Do this by organizing international conferences and workshops on the big questions that you've identified. Write persuasive commentaries for disciplinary journals and for high-impact journals, if possible. Provide service to your scientific society by seeking leadership roles on committees. These activities will help you to strengthen your academic-leadership identity.

Reframe your achievements. All of these activities will lead to further indicators and acknowledgment of your mid-career leadership: invitations to give talks at universities and keynote addresses at meetings; invitations to join review panels at funding agencies and editorial boards of journals. While you cannot necessarily trigger these yourself, your visibility as a leader in your discipline will.

You can also demonstrate your mid-career achievements in your CV by listing only presentations that are invited talks or keynotes and listing only your service contributions to professional societies and not just your membership. These changes can help to focus your mind and help to differentiate previous dossiers you submitted for early-career evaluations from those that you need for progressing to your next level of evaluation.

Getting unstuck in mid-career. You may encounter a mid-career plateau as you continue doing what you've done for years. But reflecting on the big questions in your field can be a way to re-excite your professional passions. Pursue those questions with zeal together with your graduate students and postdocs.

You might also consider a 6–12-month sabbatical, if you've not done so already. This is a crucial time for purposeful reflection when you can ponder the big questions; use a lab you might visit on your sabbatical as a sounding board for your ideas and perhaps try out some leadership activities there. The longer you go during your mid-career stage without developing leadership, in your personal science and within your collegial network, the tougher it is to begin due to inertia.

While citizenship and collegiality are still essential in mid-career as you become a leader within your department, external scientific leadership and the evidence thereof are the important mid-career target, something I did not understand all those years ago.

The best people to work with know things you do not.

Anonymous

14

RESEARCH VISITS AND ACADEMIC GROWTH SPURTS

Mark Twain in *Innocents Abroad* noted that "broad, wholesome, charitable views of men and things cannot be acquired by vegetating in one little corner of the earth all one's lifetime."[31] Almost 150 years later, I would say that this applies to science and science careers, even in our digitally connected world.

As a new lecturer or assistant professor on research visits to labs abroad, I gained exposure to new tools and new colleagues and access to high-quality PhD and postdoc recruits. Most of this exposure came about through networking (as discussed in Chapter 5) and through friendships I made along the way.

I recall spending a couple of months with a lab in Japan in the mid-1990s. Lab members were hydrologists like me, but most were trained from a completely different academic tradition from those in North America and Europe. Because most of Japan's landscape is characterized by steep, unstable slopes, most hydrologists there had backgrounds in geotechnical engineering. This produced

Navigating an Academic Career: A Brief Guide for PhD Students, Postdocs, and New Faculty, Special Publications 74, First Edition. Jeffrey J. McDonnell.

a whole different approach to field work, to modeling and to conceptualizing how hillslopes work. I marveled at their devices and at the way they thought, and I incorporated this way of thinking into my own work upon my return.

The new perspectives I gained helped accelerate my growth as a scientist. Equally important, my visit also later helped prompt a number of very high-quality, self-funded Japanese postdocs looking for a U.S. lab to come and work with me.

Here are other direct benefits of traveling to other labs.

Helps you to communicate your research. Explaining your work in different scientific contexts and cultures helps you to develop clarity in your ideas and in your writing. Receiving questions, feedback, and perspectives that you would not have had gotten at home is immensely helpful. Again, as Mark Twain would say, research visits are "fatal to narrow-mindedness,"[32] they promote a scientific humbleness in your exposure to other approaches, to other ways of observing, to other ideas and ways of thinking.

Helps you recruit PhD students and postdocs. I've recruited some of my best postdocs after spending time with them as PhD students at their home institution, selling them on the idea of joining my lab after their local graduation. I hooked one top postdoc when he was in the early months of his PhD at a university in Germany that I was visiting. I checked in with him a couple of times per year, reminding him (half jokingly and half seriously) that he needed to "fulfill his destiny" and come to me for a postdoc. Five years later, he followed up on our jocular agreement. His postdoc was a two-year scientific thrill ride of new learning, for me especially. Now, he is the leading German professor in his field, and I am the lucky one to have had him, and his exceptional mind, for two years in my lab. All of this was the consequence of my three-week summer visit to his university. I've recruited many other PhD students and postdocs this same way.

Brings exposure to new ways of scientific practice. Research visits, especially to foreign labs, accomplish more than the obvious and tangible. They help round out an early-career scientist, exposing them to new ways of working. In some cultures, perhaps a morning and afternoon group coffee is mandatory; tools and perhaps methods and approaches may be unique to that lab or country. My best visits have been with colleagues who are doing related work but are from different backgrounds and training, I as a geoscientist and they as an engineer, ecologist, or geomorphologist. The business world knows this: "The best people to work with know things you do not."[33]

Offers opportunities for visits and networking. For early-career scientists especially, academic visits afford great networking opportunities. They can even generate exposure to new research funds through your connections to other countries' national science programs or as a collaborator with your host and their groups.

Supports your thinking. Visits can also be a quiet time for reflective thinking. Some of my best ideas have come from being in a foreign land, observing or thinking about some hydrological process and having the undistracted time to think it through and develop the idea. I believe that travel helps to open minds and that research travel is the best medicine for stuck thinking. Seeing landscapes outside my home area as a field hydrologist has helped me consider processes and patterns that I would not have contemplated/pondered/thought about at home. Analogical reasoning can flourish in a new geographical setting, regardless of your scientific specialty. An extra bonus of travel is the seeding of new research ideas while you're there, ideas that, if you're lucky, your host lab might run with. Long after you have left your host lab, that group's continued pursuit of your ideas can turn into collaborative papers.

Promotes the independence of your home team. Your absence from your home lab is not necessarily a bad thing for your team. It can push graduate students to figure things out for themselves and to use peer-to-peer mentoring within your group to maintain productivity and *esprit de corps*. Weekly sessions on Skype or other platforms can work well for one-on-one sessions with PhD students and postdocs; admittedly I personally find lab group meetings more difficult to conduct from away.

Boosts your visibility. In early career, visiting key figures in your field and building meaningful collaborations with them can trigger many benefits: invitations to join a journal's editorial board or an international research initiative that they may be leading, or perhaps an offer to join in on some multiauthored vision paper they are working on. These visits and their useful consequences get you and your work noticed in a way that published papers and short conference presentations rarely provide. The visits give these key luminaries a chance to get to know you as a person, whether over dinner, on a weekend outing or in any similar setting. My most meaningful and longstanding collaborations have been the result of such visits (either my visiting them or their visiting me), and friendship helped catalyze ongoing mutual interest in collaboration.

For my past academic visits, single days have been good, weeks have been better, and a month or more better yet. Not every year affords such a luxury, but making these trips a high priority and always being on the lookout for such opportunities is one way to make them happen. Doing an academic visit takes active planning. A simple way to start is by tacking one on to a conference trip, maybe in the summer, or during a long holiday or break between terms.

Funding for research visits is usually easier than you might think. Many countries have local and national programs to bring you there. The U.S. National Science Foundation's International Program has previously had significantly higher odds of funding success, especially for some countries.

My enthusiasm for the promotion of research visits is tempered by the scientific community's increasing recognition of the impacts of travel on CO_2 emissions. Flygskam,[34] what the Swedes call "flight shame," and the admission of the effects of flying on CO_2 emissions are real. If a train option exists, I certainly take it. I also try to avoid short flights, since takeoff and landing are the real CO_2 culprits. I also try to emulate climate scientist Katharine Hayhoe at Texas Tech

University, who addresses the issue by "collecting her invitations, organizing them carefully, recommending a local scientist or requesting a video talk wherever possible (about 90% of the time!)."[35] When she does travel, she tries to do as many talks as possible in the same location. Of course, there are also carbon offset programs within and outside of the airlines. On the home front, I ride my bike to work daily, even at −30 °C with my studded tires. Perhaps that's my CO_2 self-flagellation for all this!

At the end of the day, we all need to travel less. But when we do travel, we can be more targeted and strategic (and many great labs to visit are only a train ride away). And from what I have experienced, academic visits can pay clear dividends, especially in early career. It can be helpful for tenure as internationalization is the rallying cry at campuses worldwide and in university mission statements everywhere. It will renew your level of appreciation for your own home institution. Or, if you return feeling the opposite, then this too will be a useful perspective on what you might strive for in the future with another position elsewhere.

15

ON LEADERSHIP

As I was training my first cohort of PhD students and postdocs, I focused on making them technically competent. As time progressed and after some of the mid-career bruises I experienced (as outlined in Chapter 13), I realized that, in addition to turning out technically competent people and perhaps followers of my science, I needed to turn out future leaders.

The scientific enterprise needs leadership from researchers to advocate for science and evidence-based, informed decision-making. For that, we need a wave of early-career leaders.

Early-career scientists often think about leadership in a formal sense and at the highest levels, an appointed administrator at a university, an elected leader of a scientific union. But leadership usually starts informally, and for it to take root

Navigating an Academic Career: A Brief Guide for PhD Students, Postdocs, and New Faculty, Special Publications 74, First Edition. Jeffrey J. McDonnell.

early in a career, we need to teach and mentor it as we do other acquired skills. But unlike a master's of business administration program, most U.S. PhD programs do not provide leadership training.

In discussions with my former Dean, we have found that leadership skills are cultivated over time. Incremental opportunities for increasing and diversifying leadership experiences are key for junior researchers to gain confidence, express ideas publicly, and take a stand. Universities and scientific societies do not usually do a good job of cultivating young leaders and developing the scientific enterprise's collective capacity for such leadership. This needs to change. We each found our early leadership pathways through senior colleagues who saw in us a willingness to speak up and volunteer. They mentored us. But such grooming of early-career leaders is often the exception rather than the rule. In our roles as deans at a university and presidents of large professional scientific societies, we have tried to groom early-career leaders through several paths:

Selecting junior colleagues for speaking engagements or as panel members to represent the department or university unit, so that they learn how to speak from that perspective rather than from only the point of view of their own research or self-interest.

Populating society technical committees with PhD students and postdocs to give them insight into organizational practices.

Encouraging new faculty members to serve on departmental or other committees to learn about hiring practices, merit, and tenure standards.

Presenting leadership opportunities as just that, opportunities, not unwanted service obligations.

Soliciting opinions from early-career scientists and encouraging them to speak up in departmental and professional society meetings.

Including early-career scientists on society awards committees to show them what constitutes scientific excellence and how it is rewarded and to help them to form opinions about it.

Encouraging postdocs to participate in faculty meetings to learn how university politics work and how decisions are made and some of the constraints that limit action.

Senior faculty members are busy and often overlook the potential for junior researchers to contribute. These mindsets limit our ability to see and shape young leaders. Leadership training can help younger colleagues to feel more confident in many roles, such as channeling the concerns of muzzled government scientists to the media; leading action groups within professional societies to take principled stands; chairing important committees and task forces that demonstrate the value of science; and articulating the value of evidence-based decision-making and the perils of fake news. These strengthen scientists as leaders and leaders as scientists.

Senior faculty members can mentor junior colleagues and help them to see that leadership begins in small, informal roles. As an academic community, we need to become more articulate, persuasive, and influential in holding up the values of science and the leadership that this requires, in all its forms. Grooming the next generation of leaders is a crucial step towards this goal.

An academia career is like a pie-eating contest where first prize is more pie.

SAS@academics.com

16

THE SUSTAINABLE PROFESSOR

The early years of my first university faculty position were rough. That was not because of problems, exactly, but because of opportunities, too many of them. I did not know how much was enough, so I just did more and more. As a result, I lived a life distracted, both at home and at work, with too much to do and too many people to possibly satisfy. Guilt was a constant companion, for not spending enough time with my family, for not devoting enough time to my students, and for not accepting a review request or committee assignment. It simply was not sustainable. It took me several years after getting tenure to come back to some semblance of a balanced life.

Now, when I mentor early-career scientists, I warn them about the unsustainability trap that I fell into. And I try to instill in them the idea that the goal is to stride across the finish line, whether they are completing a postdoc, getting tenure,

Navigating an Academic Career: A Brief Guide for PhD Students, Postdocs, and New Faculty, Special Publications 74, First Edition. Jeffrey J. McDonnell.

or reaching some other career goal, with a smile on their face, not in a state of collapse.

But how can you do this yourself? A sustainable professor is still a hard-working professor. Combining hard work with laser-like focus and ruthless time management is an important step toward making your life sustainable. Even more important is opportunity management.

Early on, I worried that each new opportunity, an invitation to give a talk, participate in a proposal evaluation panel, or join a committee of a scientific society, might never come along again. I felt like I couldn't say no. I now understand that early-career opportunities, like gray hairs, don't stop appearing and that sometimes it's important to turn them down so that you can complete the things you've already said yes to.

As you work on learning to say no, the "want to–need to" matrix can be a useful tool: Say yes only to the things you both need and want to do, and say no if you either do not need or do not want to do something.

Similarly, developing a personal work philosophy can help you allocate your most precious resource: time. "Work on important problems" is a good adage for starters. Beyond that, your research agenda may help you set limits. Maybe you are interested only in interdisciplinary research, or in projects that strike some balance between basic and applied research, or in pure research in a very narrow area. You can also use fun as a meter for accepting new projects or engaging in new collaborations, as I try to do. Regardless of your specific priorities, thinking through these types of questions early on can help you know when to say yes and when to politely decline.

Ultimately, self-reflection is key to sustainability. Learn to recognize the warning signs of overload. If you are in a bad place, taking stock is an essential step toward getting back on course. I wish I had realized earlier the value of a midday gym visit to mentally reset and bust stress before it accumulates. For my first several years, I could not imagine indulging in what felt like such a luxury, but taking time out pays worthwhile dividends.

Even the sustainable professor will still have days, or weeks, when their nose might slip below the water line. But more often, you will feel generally on top of things and satisfied with the direction in which your work and life are going. You'll notice that you're in the moment and not distracted in meetings or at home at the dinner table. You'll sleep well. You'll make your weekends mini-vacations and take your full allotment of vacation days. An improved home life will enhance your work, boosting job satisfaction and productivity. And when opportunities come knocking, you'll know whether to open the door.

17

A SCIENTIST LOOKS AT 60

Sixty seemed an impossibly long way away when I was an early-career scientist. I was consumed by the immediate with little time for reflection. I've been luckier than most, steeped in luck, I might say: good health (knock on wood); two now-adult happy and successful kids; and some modest career success. I'm at an age and stage where I want for nothing more, all future time seems like gravy, or "bonus time," as one of my close colleagues likes to say.

What should I tell readers like you, who are PhD students and postdocs, about the road to 60?

Perhaps that it is a marathon (sometimes wrapped inside an Ironman competition, combined with a few 100-meter-dash competitions and a relay). But it's a marathon nonetheless. Therefore, you should plan to be running with the same people for a very long time, so being nice to them and treating them well is essential.

Navigating an Academic Career: A Brief Guide for PhD Students, Postdocs, and New Faculty, Special Publications 74, First Edition. Jeffrey J. McDonnell.

You'll have a student support crew along the way. Treating them well and appreciating them is also key, as is helping them become marathon runners who know in advance that marathons are really hard work. But the satisfaction of completing the marathon to the best of your ability is enormous.

But this is not a story of a journey's end: I still have at least 10 kilometers to go!

My PhD students and postdocs find my early-career days thoroughly foreign. I am perhaps a fossil record of that change: I'm part of the last cohort of PhD students pre-Internet: e-mail and cell phones came only in my first years as an assistant professor. Back then, faculty positions that were advertised in my field drew a dozen applications at best. Graduating with a PhD meant a faculty position almost *ipso facto*. Postdocs were rare. Professional societies were small and run largely by volunteers. Meetings were small with broad sessions and mixed and varied participants. Back then, libraries housed journal hard copies and were filled with finite scientific resources. People read more and published less.

Today, science happens at the speed of light. The ratio of PhD holders to assistant professor positions gets worse each year; job ads regularly net more than 100 applicants for each. There is a worldwide glut of postdocs. In Canada, recent data suggests that "a third of graduates with doctoral degrees from Ontario universities are in tenure-track positions somewhere in the world, with half of them working as professors in Canada,"[36] not so *ipso facto*! Success rates on research proposals have dropped. My last U.S. National Science Foundation proposal was one of about 130 expressions of interest in an interdisciplinary competition: three were funded. Scientific societies like mine, the American Geophysical Union, run annual meetings with more than 25,000 participants, and the organization is fully corporate, with 100 plus employees. Ultra-narrow technical sessions are now the norm.

And publishing. Don't get me started! People read less and publish more, everyone now seems to have the Benjamin Disraeli mantra: "Whenever I want to read a paper, I write one!"[37]

So what does all this mean for today's young scientist?

The forces that are changing the academic career are indeed tectonic. But this is not a story of a sexagenarian raging against modernity. It is perhaps a recollection of a kinder, gentler science practiced pre-Internet, in which all science was slow (now we have "slow science movements"[38] that harken back to these times). Don't get me wrong. I love being a scientist today. And despite the intensity, each day is a thrill ride of discovery, about myself and about my field.

But perhaps the most important thing for early-career folks today is that life moves very quickly. I discovered the moral essay "On the Shortness of Life" by Lucius Seneca (a Roman philosopher and statesman) about 15 years ago. It's an exceptionally relevant discussion of not wasting time. His take was, "It is not that we have a short time to live, but that we waste a lot of it."[39] "Don't waste time" might be the most important single message in this book too.

Other things that I have observed to be passed down: that it is a Long Game. Short gains (a *Nature* paper, a huge grant, a major talk) will generate interest only in the year that they were awarded, after that, it's back to "What have you done for me lately"? Know this ahead of time, to avoid disappointment or unrealistic expectations. Chapter 9 discusses the perils of roaming too far and too widely. Stay focused and on course, there are so many forces to drive you off course.

Being a scientist today can be much like being a research manager. Avoid this pitfall as much as possible. Carve out a precious hour or two every day for thinking and writing and for your own ideas. This is what will sustain you at 60. Of course, working on important questions and with the best people you can find, as students and collaborators, is crucial. Unimportant, but perhaps highly fundable, questions are like empty calories. The important ones will sustain you nutritionally. So too will your students and postdocs sustain you, as will watching their careers (and their students' careers) unfold, even when you are in your scientific rocking chair.

Looking back, I realize now that my expectations were very low going in; I had none. And as the first in my family to go to university, I had no frame of reference. Everything seemed like some sort of club I somehow sneaked into. My imposter syndrome in the early years was extreme. So, my happiness with all this was (and is) high. So, know that happiness = reality − expectations.[40]

Having a life philosophy can help. For some, this will come from a religious tradition; for others, perhaps it can come from their early immersion in classic literature. For me, it has been Jimmy Buffett. If Jim Morrison was the Lizard King, Jimmy Buffett is the Philosopher King. His music is an encapsulation of the Robert Louis Stevenson's writing, "To miss the joy is to miss all." Just listen to a few of his songs: A Pirate Looks at Forty; Changes in Latitudes, and Changes in Attitudes; Trying to Reason with Hurricane Season; I Have Found Me a Home; and One Particular Harbor. I dare you to listen to these songs and not feel his life philosophy take hold of you!

Like many, I am gripped by the need to explain, in my case, the natural world around me and, in particular, how water cycles through it. I imagine that retirement, when it comes, will be more akin to a permanent sabbatical. But even now, I am slowly starting to shift my personal balance from research to mentoring. Nevertheless, I live with the hope that my best papers are yet to come. I am well aware of the old cartoon tombstone: "Published, but Perished Anyway." Most of all, I am appreciative for this bonus time. I hope to stay young while getting old. Who knows if I will succeed, ask me when I am 100!

The things which are
most important don't
always scream
the loudest.

Bob Hawke, 23rd Prime Minister of Australia

POSTFACE

I've tried in this book to highlight some of the key issues before, during, and after the tenure years. You will notice that there is no mention of citations, h-index, and the like. That is because, in the end, these things can needlessly distract the early-career scientist. My advice in this regard is simple: work hard, do good work, and be a good scientific citizen. The various metrics out there have become something of a digital water torture for many. Ignore these as much as possible.

By no means are these reflections and advice comprehensive. Rather, I've aimed with them to address some of the unwritten rules of the game. It is a wonderful game, to be played and enjoyed.

Academia can bring some of life's greatest joys: watching a PhD student submit, defend, and go on to an academic job and thrive in it. Ah, let me savor that sentence for a while. And, academia can bring some of life's greatest stresses,

Navigating an Academic Career: A Brief Guide for PhD Students, Postdocs, and New Faculty, Special Publications 74, First Edition. Jeffrey J. McDonnell.

a monster project that colleagues have encouraged you to lead; getting word of funding success and then realizing the full horror that awaits you for the next three years.

It is also important to remember that there is more to life than academia alone. There is more to life than your job. The famous Harvard University psychologist Erik Erikson said that "the richest and fullest lives attempt to achieve a balance of work, life and love ... for a life of achievement and serenity." That's the trifecta to seek.

We hear so much about work–life balance these days. I'm not sure that such a thing exists in academia (remember: always working, never working, in Chapter 1). Maybe in the end, it is as simple as Michael Pollan's "Omnivore's Dilemma,"[41] "Eat food, mostly plants, not too much," where the academic's dilemma is Work hard, not too much, mostly on papers.

I recall a great tweet on SAS@AcademicsSay (a must-read site!) that "academia is like a pie-eating contest where first prize is more pie." You survive this with your values: values you can choose and values that keep you going through the ups and downs of academic life. As many papers in psychology attest "Values get you to what is enough."[42] Values save you from the pie-eating contest. They let you be you and not succumb to comparisons with others.

I'll end with a quote attributed to U.S. President Dwight D. Eisenhower: "What is important is seldom urgent and what is urgent is seldom important." This captures academic life in a nutshell: the tyranny of the trivial. Academic life is best done with earmuffs on (things we wear in Canada in winter), to drown out the white noise of what is coming across your desk and being asked of you, and to filter out the unimportant. It's daily triage, but with a long-term goal and purpose. Losing track of that long-term goal, that *one thing*, is so easy to do and, sadly, so common in academia today. We float like flotsam and jetsam from funded project to funded project, forgetting what exactly we want to do. While this can be satisfying for some, it robs the world of your deeper impact and bars you from moving the dial on something that can be taken up by the next generation. I hope that some of these chapters can help counterbalance some of these forces.

ACKNOWLEDGMENTS

I must first thank my family for their support of (and participation in) my academic journey. My father, John, a policeman, taught me the value of hard work; my mother Goldie, a stay-at-home mom (and "Tiger Mother," before it was a thing), taught me focus. My parents are the headwaters of my luck stream. My father grew up in subsidized housing in the United Kingdom and my mother on a small farm in Ontario. I remember them accompanying me to a University of Toronto open house in my final year of high school. The three of us walked around like we'd been parachuted into outer space.

My children, John and Meghan, grew up in an academic household in a U.S. college town. Their plight was perhaps the opposite of mine, outer space was their

---

Navigating an Academic Career: A Brief Guide for PhD Students, Postdocs, and New Faculty, Special Publications 74, First Edition. Jeffrey J. McDonnell.

natural habitat! They are my two most import mentees and I am lucky to be their sounding board as their adult lives unfold. I thank my wife, Veva, for being a loving, constant companion and steady supporter of my work (and for her edits of this book). Together, my family are the ones who keep me focused on "living for my eulogy and not my CV."[43]

I must thank my mentees: my past graduate students and postdocs, and academic visitors to my lab. I am sure that I learned more from them than they learned from me. Most of all, I appreciate their tolerance of my mentoring efforts, successful and unsuccessful. I really was inept in the early years. I hope that I am better now. I am still learning with each student and each passing year.

Rituparna Bose from Wiley and Brooks Hanson and his team at AGU are thanked for their enthusiastic support along the way. I thank Laurent Pfister and the Luxembourg Institute for Science and Technology (LIST) for sabbatical support during the writing of this book. And I thank the LIST PhD students and postdocs for serving as critics on earlier versions of this text and especially Wouter Berghuijs and Nicolas Rodriguez for useful critical feedback.

Chapters 6–11 and 16 are re-edited reproductions of "Working Life" articles in *Science* magazine. Rachel Bernstein is thanked for her initial editing magic on these, each reproduction is used with permission. Chapter 8 is a partial reproduction of a paper in *Astronomy and Geophysics* with Seth Stein, Deering Professor of Earth and Planetary Sciences at Northwestern University, and M. Meghan Miller, President of UNAVCO. Chapter 15 is a partial reproduction of a *Nature* Careers article coauthored with Toddi Steelman, Nicolas School of the Environment at Duke University. Karen Kaplan is thanked for her masterful edits on this and on a version of Chapter 12. I thank these publishers and my former co-authors for their permission for reproducing that work here.

I thank Donn Forbes for unlocking the secrets of the proposal writing cabinet. His kindness and patience with me is deeply appreciated and I encourage any young faculty members who are embarking on a large grant proposal to consider his proposal writing services at donnforbes.com! I thank Kim Janzen for her early, useful edits and Karen Kaplan for kindly agreeing to a final edit of the overall book. The artwork is reproduced with permission from Robert Neubecker. He and his agent, Jodie Hein, are thanked for their release of these works.

I want to thank the 1000+ PhD students and postdocs whom I have given versions of this book to in short course form, starting in 2008 at Oregon State University, then at AGU and EGU meetings, and then at dozens of universities in Canada, United States, China, Australia, South America, and Europe. Some of the content of this book has arisen from their questions and reactions along the way. Thank you, students, I continue to learn from you.

Lastly, I want to acknowledge Don Siegel at Syracuse University, who first inspired me to think about mentoring 20 years ago. Many of my early ideas were simply his, repackaged through my own lens. Don is an incredible force of nature and is someone who mentored me about how to mentor without him knowing it. For the past six years I have co-taught portions of this book locally with Maureen Reed from the School of Environment and Sustainability. I thank her for her great collegiality and for useful discussions along the way.

If I had more time, it could have been shorter.

Blaise Pascal, 1623–1662, French mathematician

NOTES

Navigating an Academic Career: A Brief Guide for PhD Students, Postdocs, and New Faculty, Special Publications 74, First Edition. Jeffrey J. McDonnell.

1 Some have described administration in this way. I too think it is quite apt. I spent two years in an administrative role. It was like real work with most days full of back-to-back meetings. It also was like seeing how sausage is made. Two years was more than enough to convince me that the real dream job in academia is within the professoriate.

2 I am without direct experiences or study of racism and sexism. Diversity, inclusion, and implicit bias in academia are crucial issues, but are not the focus here. Within my main professional society, the American Geophysical Union, we have tried to make some headway on implicit bias with new and useful materials here: https://eos.org/agu-news/getting-to-fair-recognizing-implicit-bias-and-easing-its-impact. And within my own section, I addressed head on some of these lingering issues (see pages 1–4 in https://sfg.agu.org/hydrology/wp-content/uploads/sites/19/2017/07/HS-July-2017-Newsletter_20170712_FINAL.pdf).

3 My pre-PhD job experiences included church janitor, hospital dishwasher, piano and accordion teacher, and warehouse shipper-receiver. As a graduate student, I worked as a live-in college don (resident assistant). That gave me some insights into dealing with undergraduates, albeit mostly the intoxicated ones on weekends.

4 Engaged workaholic. I am no expert in this area but when I read a piece like this one (https://workingclassstudies.wordpress.com/2012/06/11/working-hard-or-hardly-working/), the idea that engaged workaholics "work because work is fun" really resonates. This piece and many referenced within it suggest that engaged workaholics suffer "less burnout, and perhaps fewer health deficits" than the standard workaholic. As for me, time will tell!

5 *The Research is to See* quote is attributed to Albert Szent-Györgyi (https://en.wikipedia.org/wiki/Albert_Szent-Gy%C3%B6rgyi) who first isolated Vitamin C. I love the spirit of this quote. And as a 4-g-per-day Vitamin C taker, Linus Pauling (double Nobel prize winner) is also my hero (and, fun fact: a graduate of Oregon State University).

6 Wilson, E. O. (2013). *Letters to a young scientist*. New York, NY: W. W. Norotn & Company. Retrieved from https://books.wwnorton.com/books/detail.aspx?ID=4294972234

7 Definition here: https://dictionary.cambridge.org/dictionary/english/indomitable

8 Beveridge, W. I. B. (1950). *The art of scientific investigation*. New York, NY: W. W. Norton & Company.

9 Hamming, R. (1986, March 7). *You and your research*. Transcription of the Bell Communications Research Colloquium Seminar (as posted by J. F. Kaiser, Bell Communications Research, Morristown, NJ). Retrieved from https://www.cs.virginia.edu/~robins/YouAndYourResearch.html

10 Medawar, P. B. (1980). *Advice to a young scientist*. London, England: Harper & Row.

11 Medawar, P. B. (1980). *Advice to a young scientist*. London, England: Harper & Row.

12 Frank, R. H. (2016). *Success and luck: Good fortune and the myth of meritocracy* (p. 187). Princeton, NJ: Princeton University Press.

13 Zacharia, F. (2015). *In defense of a liberal education*. New York, NY: W. W. Norton & Company.

14 Levecque, K., Anseel, F., De Beuckelaer, A., Van der Heyden, J., & Gisle, L. (2017). Work organization and mental health problems in PhD students. *Research Policy*, 46(4), 868–879.

15 Beveridge, W. I. B. (1950). *The art of scientific investigation*. New York, NY: W. W. Norton & Company.

16 Bacon's essays can be found online here: http://www.literaturepage.com/read/francis-bacon-essays.html. He also went on to say: "Reading maketh a full man; conference a ready man; and writing an exact man!" Of course, we would use a more gender-neutral pronoun today.

17 Manson, M. (2016). *The subtle art of not giving a fu*k: A counterintuitive approach to living a good life* (1st ed.). New York, NY: HarperOne.

18 Darwin, F. (Ed.). (1887). *The life and letters of Charles Darwin, including an autobiographical chapter*. London, England: John Murray.

19 Sow a thought: Usually attributed to Emerson but likely this is Biblical in original, https://en.wikipedia.org/wiki/Charles_Reade

20 Dalio, R. (2017). *Principles: life and work* (p. 592). New York, NY: Simon & Schuster.

21 Olson, R. (2015). *Houston, we have a narrative*. Chicago, IL: University of Chicago Press, p. 256.

22 Merton, R. K. (1968). The Matthew effect in science. *Science*, 159(3810): 56–63.

23 Rigney, D. (2010). *The Matthew effect: How advantage begets further advantage* (p. 176). New York, NY: Columbia University Press. But recent work described in *The Economist* (https://www.economist.com/science-and-technology/2019/05/11/in-science-grit-counts-as-well-as-talent) suggests that there is hope, even for those who do not get early advantage. This work shows that grit and determination can trump initial advantage.

24 Levine, A. G. (2015). *Networking for nerds: Find, access and land hidden game-changing career opportunities everywhere* (p. 256). Hoboken, NJ: Wiley-Blackwell.

25 Rigney, D. (2010). *The Matthew effect: How advantage begets further advantage* (p. 176). New York, NY: Columbia University Press.

26 Carnegie, D. (1936). *How to win friends and influence people*. New York, NY: Simon & Schuster.

27 National Academy of Sciences, National Academy of Engineering, and Institute of Medicine. (2009). *On being a scientst: A guide to responsible conduct in research* (3rd ed.). Washington, DC: The National Academies Press. https://doi.org/10.17226/12192

28 Olson, R. (2015). *Houston, we have a narrative* (p. 256). Chicago, IL: University of Chicago Press.

29 Donn Forbes at donnforbes.com

30 Conrad, J. (1902). *Heart of darkness*. See https://en.wikipedia.org/wiki/Heart_of_Darkness for publishing details.

31 Twain, M. [1835–1910]. (1984). *The innocents abroad; Roughing it*. New York, NY: Library of America.

32 Twain, M. [1835–1910]. (1984). *The innocents abroad; Roughing it*. New York, NY: Library of America.

33 Best people to work with: unknown reference.

34 Flygskam: *Forbes* magazine Op-ed piece by David Banis titled "Will 'Flygskam' (Or flight shame) be the buzzword of this year's summer holiday?" 26 April 2019.

35 http://katharinehayhoe.com/wp2016/

36 Chiose, S. (2016, January 24). More PhDs finding jobs as tenure-track professors, study says. *Toronto Star*. This article from *The Atlantic* https://www.theatlantic.com/business/archive/2013/02/how-many-phds-actually-get-to-become-college-professors/273434/, while slightly more dated, seems to support generally the Ontario

analysis, but with variability across different fields of study and different countries. Indeed, the U.K. stats I have seen suggest that incredibly low odds in landing a faculty position (some analysis suggesting less than 5%). And the opposite in China where conversations in recent weeks and months with several college Deans there suggesting that, at least in the environmental sciences, nearly all PhD graduates who wish to become faculty members in China's 2900 universities, can. In my own experience, I have graduated ~20 PhD students. Half of my former PhD students (my new colleagues) are now professors. A quarter of the others chose immediately to go into consulting or government research because that was what they had wanted from the outset. Another quarter are in such positions after trying their hand at obtaining a faculty position. Of my ~30 former postdocs, three-quarters of them are now professors. One-quarter are in government research (with one or two in consulting). So, whatever the odds, knowing that a PhD is a necessary, but not sufficient condition, for obtaining a faculty position is important. And hopefully, some parts of this book show the why and how of this.

37 https://en.wikipedia.org/wiki/Benjamin_Disraeli

38 http://slow-science.org/

39 Seneca (2005). *On the shortness of life: Life is long if you know how to use it*, (2005 ed.). London, England: Penguin Books.

40 *New York Times* op-ed piece by Tim Herrera, A deceptively simple way to find more happiness at work, 7 April 2019.

41 Pollan, M. (2006). *The omnivore's dilemma: A natural history of four meals*. (p. 450). New York, NY: Penguin Press.

42 Article in *Psychology Today* by Steven Hayes, 12 September 2018, https://www.psychologytoday.com/us/articles/201809/10-signs-you-know-what-matters

43 *New York Times* op-ed piece by David Brooks, The moral bucket list, 11 April 2015.

Chapters 6–12 and 15–16 are sourced from original materials here and used with permission by the publishers:

McDonnell, J. J. (2019). Step up to leadership for mid-career growth. *Nature*. doi:10.1038/d41586-019-01936-7

McDonnell, J. J. (2017). The sustainable scientist. *Science, 357*(6356), 1202. doi:10.1126/science.357.6356.1202

McDonnell, J. J. (2017). Paper writing gone Hollywood. *Science, 355*(6320), 102–102. doi:10.1126/science.355.6320.102

McDonnell, J. J. (2016). Orchestrating a powerful group. *Science, 352*(6283), 378. doi:10.1126/science.352.6283.378

McDonnell, J. J. (2016). The 1-hour workday. *Science, 353*(6300), 718. doi:10.1126/science.353.6300.718

McDonnell, J. J. (2015). Creating a research brand. *Science, 349*(6249), 758. doi:10.1126/science.349.6249.758

Steelman, T., & McDonnell, J. J. (2017). Look for the leaders. *Nature, 547*, 483. doi:10.1038/nj7664-483a

Stein, S., McDonnell, J. J., & Miller, M. (2018). Discussing scientific ethics: What would you do? *Astronomy and Geophysics, 59*(4), 4.12. doi:10.1093/astrogeo/aty187

FURTHER READING

There is a lot of good material out there to read! The "Working Life" section of *Science*, edited by Rachel Bernstein, is a great start. These weekly pieces offer personal insights into what it is like to launch a science career today. There is also the Careers section of *Nature,* edited by Karen Kaplan, with many varied pieces on academic and research life. Many of these deal with crucial issues in academia and with many aspects that are not discussed in this book: gender, diversity and inclusion, mental health, discrimination, harassment.

Many books are listed below as further reading. All are useful. For writing, Randy Olson's book, *Houston, We Have a Narrative*, is a classic. I discovered it after writing "Paper Writing Gone Hollywood." He discusses the ABT (the "and, but, therefore") approach to storytelling that is so very helpful for the academic

Navigating an Academic Career: A Brief Guide for PhD Students, Postdocs, and New Faculty, Special Publications 74, First Edition. Jeffrey J. McDonnell.

elevator pitch. One of the best for understanding how academia works is "Academic Duty" by Donald Kennedy, former Editor-in-Chief of *Science* magazine and former President of Stanford University. It is a masterpiece on what the modern university career is all about. The U.S. National Academies report "On Being a Scientist" is exceptionally good. A timeless classic is William Ian Beardmore Beveridge's 1950 book *The Art of Scientific Investigation*. If one reads only one book (besides mine of course!), then Beveridge's would be it. Lastly, many great biographies and autobiographies of scientists exist. As a Canadian, I am partial to David Suzuki's first autobiography from 1987, *Metamorphosis: Stages in a Life*. Perhaps the best of all time is Darwin's autobiography. Published 100 years before Suzuki's, it provides enormous insight into analogical reasoning and the source of creativity in science. His passion quotient and creativity quotient were "off the chain" as a student today might say today.

Here's my curated list.

Belcher, W. (2009). *Writing your journal article in 12 weeks: A guide to academic publishing success*. London, UK: SAGE. 351 pages, ISBN: 141295701X, 9781412957014

Beveridge, W. I. B. (1950). *The art of scientific investigation*. New York, NY: W.W. Norton & Company.

Boice, R. (2000). *Advice for new faculty members: Nihil Nimus*. Upper Saddle River, NJ: Pearson. 336 pages, ISBN: 0205281591, 9780205281596

Chakrabarty, P. (2012). *A guide to academia: getting into and surviving grad school, postdocs, and a research job*. Chichester, UK: Wiley-Blackwell. 192 pages, ISBN: 9780470960417

Craufurd, D., & DeNeef, A. (2006). *The academic's handbook*. Durham, NC: Duke University Press. 416 pages, ISBN: 0822388200, 9780822388203

Feibelman, P. (2011). *A PhD is not enough!: A guide to survival in science*. New York, NY: Basic Books. 160 pages, ISBN: 0465022227, 9780465022229

Fiske, P. S., & Louie, A. (2001). *Put your science to work: The take-charge career guide for scientists*. Washington, DC: American Geophysical Union. 179 Pages, ISBN: 9780875902951

Gray, P., & Drew, D. (2012). *What they didn't teach you in graduate school: 299 helpful hints for success in your academic career* (2nd ed.). Sterling, VA: Stylus Publishing, LLC. 227 pages, ISBN: 1579226434, 9781579226435

Hamming, R. (1986, March 7). *You and your research*. Transcription of the Bell Communications Research Colloquium Seminar (as posted by J. F. Kaiser, Bell Communications Research, Morristown, NJ).

Kennedy, D. (1999). *Academic duty*. Cambridge, MA: Harvard University Press. 320 pages, ISBN: 0674002229, 9780674002227

Lang, J. (2005). *Life on the tenure track: Lessons from the first year*. Baltimore, MD: Johns Hopkins University Press. 186 pages, ISBN: 0801881021, 9780801881022

Levecque, K., Anseel, F., de Beuckelacr, A., van der Heyden, J., & Gisle, L. (2017). Work organization and mental health problems in PhD students. *Research Policy*, *46*(4), 868–879. doi:10.1016/j.respol.2017.02.008

Olson, R. (2015). *Houston, we have a narrative* (pp. 256). Chicago, IL: University of Chicago Press.
Reis, R. (2012). *Tomorrow's professor: Preparing for academic careers in science and engineering*. Hoboken, NJ: Wiley. 440 pages, ISBN: 9780780311367, 9780470546727
Smaglik, P. (2017, December 20). Know the odds. *Nature*. Naturejobs blog. Retrieved from http://blogs.nature.com/naturejobs/2017/12/20/know-the-odds/
Toth, E. (2008). *Ms. Mentor's new and ever more impeccable advice for women and men in academia*. Philadelphia, PA: University of Pennsylvania Press. 272 pages, ISBN: 0812220390, 9780812220391

INDEX

Navigating an Academic Career: A Brief Guide for PhD Students, Postdocs, and New Faculty, Special Publications 74, First Edition. Jeffrey J. McDonnell.

The author in 2019 navigating his own academic career.